A STEP-BY-STEP BOOK

ANIMALS *in* THEIR HOMES

Written by Anita Ganeri

Artwork by John Mac and John Rielly

WISHING WELL BOOKS®

CONTENTS

Introduction	5
Busy, Buzzing Hive	6
Clever Weavers	8
Living in Lodges	10
Spider in a Spin	12
Mouse House	14
Tailor-Made	16
Hills and Holes	18
Beautiful Bower	20
Treetop Retreat	22
Wasp Pots	24
A Fishy Nest	26
Fun Quiz Time	28
Glossary	30

Published by Wishing Well Books,
an imprint of Joshua Morris Publishing, Inc.,
221 Danbury Road, Wilton, CT 06897.

Copyright © 1994 Salamander Books Limited,
129-137 York Way, London N7 9LG.
All rights reserved. Printed in Belgium.

ISBN: 0-88705-755-1

10 9 8 7 6 5 4 3 2 1

INTRODUCTION

Animals build their homes in a variety of places —
underground, underwater, high up in the trees. These homes
vary from nests and burrows to complex honeycombs and
lodges. But however they may differ, most animal homes
provide their owners with a safe, warm place to rest, take
shelter from the weather, and raise their young.

Through step-by-step illustrations, this book shows how
various animals construct their homes — often with amazing
skill and a remarkable range of materials. At the end of
Animals in Their Homes, you'll find a fun quiz to test your
knowledge as well as a glossary to explain some of the
special terms used in the book.

BUSY, BUZZING HIVE

Honeybees are very clever builders. A colony of thousands of bees can build a new home, or hive, in two or three days. Different groups of bees, called worker bees, concentrate on different parts of the hive at the same time — with amazing results. The cells they build are always the same size and shape, and their walls are always the same thickness. Also, the honeycombs (made up of rows of cells) always hang in the same direction. Human architects would need all sorts of calculators, rulers, and charts to help them build so accurately.

Every now and then, a swarm of bees, led by an old queen bee, leaves its hive and sets off to build a new home. As soon as they find a suitable spot in a hollow tree, the bees begin their skillful work.

1

First, the bees cluster together to keep warm. The warmth helps them to make wax in special glands underneath their abdomens.

2

The bees scrape the wax off of the glands with their back legs. Then they mix the wax with spit and mold it into soft balls with their front legs and jaws.

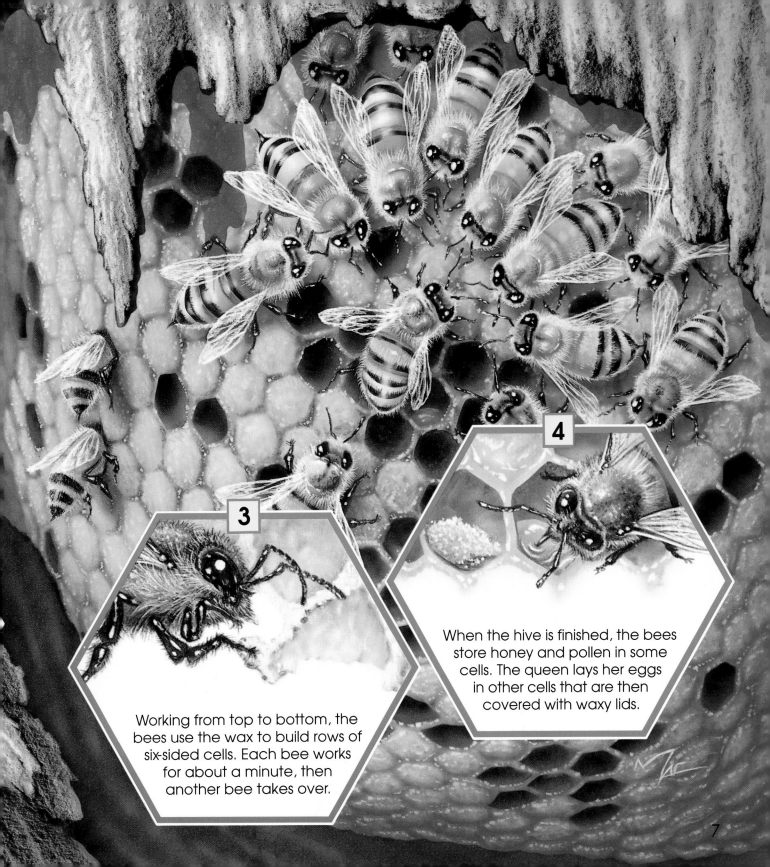

3

Working from top to bottom, the bees use the wax to build rows of six-sided cells. Each bee works for about a minute, then another bee takes over.

4

When the hive is finished, the bees store honey and pollen in some cells. The queen lays her eggs in other cells that are then covered with waxy lids.

7

CLEVER WEAVERS

Many birds build nests as safe places in which to lay their eggs and raise their young. But very few construct nests as amazing as those of the weaverbirds. The male bird builds a round nest of grass or palm fronds, which hangs like an exotic fruit from a tree branch. Using his beak and feet, he weaves the leaves together and holds them in place with cleverly tied knots. It's not as easy as it looks — young birds may have to try again and again until they finally get it right.

When the nest is finished, the male waits for a female to come and inspect his work. If she likes what she sees, she chooses his nest to lay her eggs in. Then she lines the nest with soft feathers and grass to make it warm for the eggs.

Some weaverbirds build their nests together in the branches of the same tree, with one large roof over all of them. A single tree may contain as many as 300 nests.

1

The male weaverbird tears off long blades of grass with his beak. Then he winds the grass around a forked twig and ties it in place with firm knots.

2

The bird weaves and knots more grass into the first strips to make a ring. He holds the strips steady with his feet and ties the knots with his beak.

3

As he continues to weave in more grass, the bird gradually builds up the walls of the nest until it begins to take on its final ball shape.

4

Finally, the weaverbird adds a long, tubelike entrance below the nest. This makes it difficult for snakes and other thieves to steal the eggs.

LIVING IN LODGES

Beavers can be found in and around rivers and streams. They are well suited to life in water, with webbed back feet to help them swim and large flat tails to power them along and help them to steer.

Beavers work together as a family to build their home, or lodge. They cut down trees with their strong, chisel-like teeth and use the logs and branches to construct a dam across the river. This forms a small, deep lake in which the beavers can build their wooden lodge.

The beavers build their dam out of logs, branches, sticks, and stones, which they carry underwater in their mouths or front paws. They plug any gaps in the dam with mud to make it watertight.

2

When the dam reaches about 3 feet above the surface of the water, the beavers build an island of logs, branches, and bark in the middle of the lake. This may reach 6 to 10 feet above the surface.

3

Then the beavers hollow out a living chamber inside the lodge. They reach the chamber by a series of underwater passages. Finally, the beavers cake the chamber walls with mud to keep their home warm.

SPIDER IN A SPIN

The best time to spot spiders' webs is in the early morning when they glisten with dew. Many spiders spin webs as traps for insects. These are made of silk, produced inside the spider's body and squeezed out through tiny nozzles, called spinnerets, at the end of the spider's body. The silk is soft at first, but hardens in the air. It is stronger than any other natural fiber — even stronger than steel wire of the same thickness.

Spiders make different types of silk for different uses. Dry threads are used for the main frame of the web, and as scaffolding. But the spiral is spun using sticky, glue-coated silk. Any insect that flies into the web gets stuck and cannot escape. Then the spider, which has been waiting for its meal, pounces.

Not surprisingly, the web suffers a great deal of wear and tear. The spider carries out repairs every day and has to spin a whole new web every other day.

1

First, the spider spins a thread of silk between two twigs, which acts as a bridge across the gap. Then it adds three further threads to form a Y shape.

2

The spider adds more silk spokes to the Y shape, attaching some spokes to the surrounding twigs. Finally, the main frame of the web is completed.

3

To hold the spokes in place, the spider temporarily laces them together with a spiral of dry silk. Like scaffolding, this stops the web from falling apart.

4

Then, working inward, the spider spins a spiral of sticky thread, eating the scaffolding as it goes. At last, the trap is set for an unsuspecting insect.

MOUSE HOUSE

Harvest mice are tiny rodents that live among the tall stalks of corn and grass in fields and meadows. They use their long tails for support as they climb among the plants. In summer, the female harvest mouse builds a nursery nest for her young. The nest is about the size and shape of a tennis ball. It is bound tightly onto the plant stalks, high above the ground where it is safe from the mouse's enemies.

A mother harvest mouse has several litters of babies every year. She has to make a new nest for each litter. The babies are very active and soon pull their old nest to pieces as they play and explore. Harvest mice also build winter nests underground or among tree roots. In the fall, they stock these nests with food, ready for the coming winter.

1

With her front paws, the harvest mouse pulls blades of grass through her mouth and shreds them into strips with her teeth. She leaves the blades attached to their stalks.

2

Then the mouse laces the shredded blades of grass together to form a firm base for her nest. She works quickly, using her front paws and mouth as tools.

3 More grass is woven into the framework by the mouse to build up the sides of the nest. Then the mouse adds a dome-shaped roof to the top of the nest.

4 The mouse strengthens the inside of the nest with leaves and grass. Finally, she lines it with shredded leaves, feathers, and downy seeds to make it warm.

TAILOR-MADE

Tailor ants build their oval-shaped nests from leaves that are still attached to the tree. They have a very clever way of sewing the leaves together. Instead of needles and thread, they use their own ant larvae.

The larvae make silk in special glands in their mouths. One group of ants holds two leaves together with their legs and jaws. Another group runs the larvae backward and forward between the leaves, giving the larvae a gentle squeeze to get the silk thread flowing from their mouths. Only the larvae can be used for sewing. When they change into adults, they stop producing silk.

Some tailor ants build simple nests using only a few leaves and a loose seam of silk. Others build more complex nests with several different chambers inside.

1

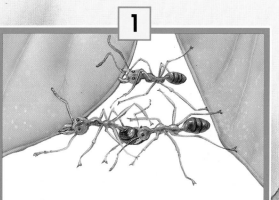

First, the ants pull the edges of two leaves together. If the gap is very wide, two or more ants hold on to each other to form a bridge between the leaves.

2

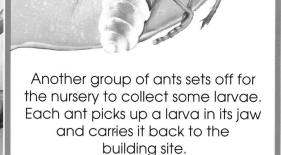

Another group of ants sets off for the nursery to collect some larvae. Each ant picks up a larva in its jaw and carries it back to the building site.

3

Then the ants pass the larvae back and forth between the leaf edges, joining the leaves together with a thick, white mesh of sticky silk threads.

4

The ants continue to pull and sew leaves together until they have completed their nest. Then, once their task is done, the ants return the larvae to the nursery.

HILLS AND HOLES

A mole spends most of its life underground. Sometimes the only signs that a mole is nearby are large piles of soil on the surface, called molehills. These are formed by the mole pushing soil to the surface as it digs its network of tunnels and chambers. The biggest molehill marks the place of either the mole's nesting chamber or its larder, which the mole keeps well stocked with earthworms.

A mole is well suited to a life of digging and tunneling. Its front paws make excellent shovels and its long nose is used to loosen the soil. The mole's short, velvety fur can bend and lie in any direction so it doesn't get caught on the tunnel walls and slow the mole down.

1

As the mole digs a tunnel through the ground, it pushes the soil sideward and backward with its large front paws, using a swimminglike action.

2

At the end of a tunnel, the mole digs out a large chamber for nesting or sleeping in. It lines the chamber with warm, dry leaves and grass.

3

Then the mole sets off to hunt for juicy earthworms to eat. It may dig a new tunnel in search of food or simply crawl along an existing tunnel.

BEAUTIFUL BOWER

Some male birds, such as peacocks and birds of paradise, show off their beautiful feathers to attract a female in the breeding season. Other birds show off their building skills. The satin bowerbird constructs a special structure of twigs, called a bower, and decorates it with flowers, berries, leaves, and shells to make it look beautiful.

The male straightens up the bower every few days and even steals decorations from his rivals when they aren't looking. If a female comes closer for a better look, the male sings and dances and picks up various objects from his collection of treasures to show her. If she likes what she sees, the two birds go into the bower and mate. The female bowerbird then builds her own nest among the trees and lays her eggs in it.

1

The male begins by clearing a patch of forest floor. Then he builds his bower by sticking two parallel rows of twigs into the ground.

2

At one end of the bower, he clears a small area of ground and lays down a carpet of grass and fine twigs. This will be his dance floor.

3

Then he decorates the bower and dance floor with bright blue and yellow objects. These may even include buttons and bottle caps.

4

Finally, the male paints the inside walls with charcoal or the juice of blue berries. He uses his beak or a piece of bark as a paintbrush.

21

TREETOP RETREAT

During the day, a family of chimpanzees roams the forest floor on the lookout for leaves, eggs, and termites to eat. The family is a close-knit group. The chimps help each other to find food, defend their patch of forest (called their territory), and to find a suitable, safe place to sleep for the night.

Each chimp builds itself a comfortable sleeping nest, high up in the trees. It has to make a new bed each night — luckily this only takes five minutes or so. The nests are made from springy branches and twigs, padded with leaves. Baby chimps share their mothers' nests. But young chimps play at nest-building from an early age, using grass and twigs instead of branches. Their first attempts don't last long, but as with most things, practice makes perfect!

1

First, the chimp chooses a firm base in the fork of a tree. Then it weaves branches and twigs into the base, holding each branch in place with its feet as it works.

2

Next, the chimp tucks in the leafy twigs growing around the edge of the nest to make it more comfortable. The chimp may even grab a few leaves to make a pillow.

3

Finally, the chimp snuggles down into the nest and presses it into shape with its body. Then the chimp settles down for a good night's sleep, safe from all its enemies.

WASP POTS

Some wasps build their nests out of paper, made from chewed up fragments of wood. But potter wasps use mud, just as human potters use clay. The females collect the mud from a damp patch of ground. If the mud is too dry, they moisten and soften it by spitting water onto it. Then they carry the mud to a bush or tree, or to a sheltered spot among some rocks. Here they build their tiny, bottle-shaped nests, either singly or in little clusters.

When a nest is finished, the female stocks it with half-dead caterpillars or spiders and lays an egg in it. The egg hangs by a thread from the inside rim of the nest. In this way, when the wasp larva hatches, it has a meal of fresh meat ready to feast on.

1

The female potter wasp molds lumps of damp mud into balls with her jaws and front legs. She carries the balls one by one to her building site.

2

Using her front legs and jaws, she shapes the mud into long, thin strips. Then she lays the strips down in rings, building up the walls of her tiny nest.

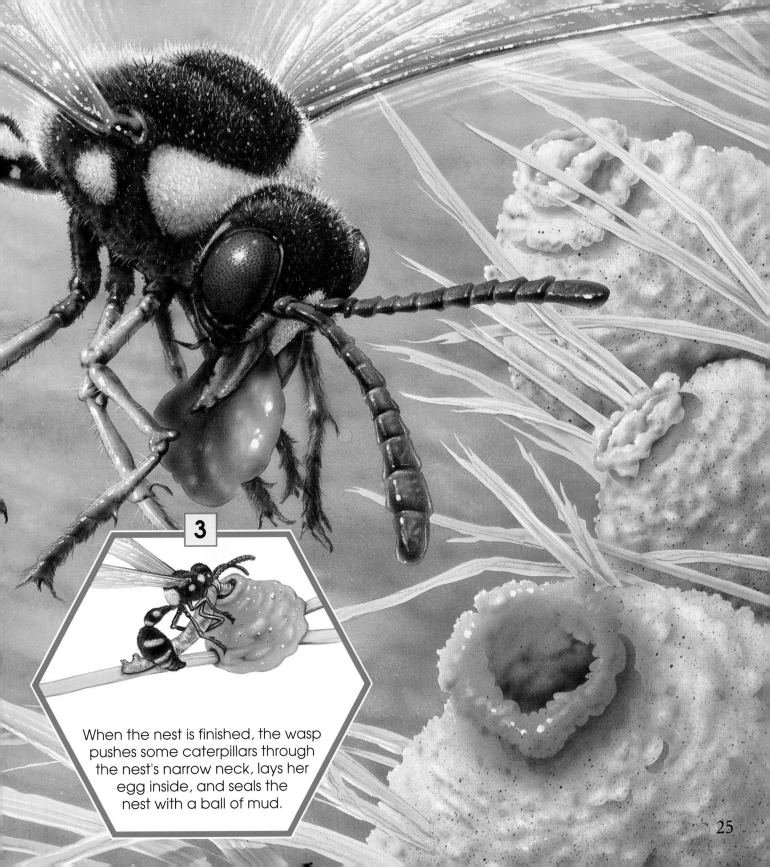

3

When the nest is finished, the wasp pushes some caterpillars through the nest's narrow neck, lays her egg inside, and seals the nest with a ball of mud.

25

A Fishy Nest

The stickleback is one of the very few fish that build nests. In spring, the male stickleback uses plant material to build his nest on the bottom of the pond or riverbed. When the nest is ready, he invites female fish to come and lay their eggs in his nest, dancing in front of them to guide them inside. Several females lay their eggs in the same nest, and there may be several hundred eggs in all. Then the male fertilizes the eggs.

The male looks after the eggs with great care, chasing away or fighting off any intruders. He fans fresh water onto the eggs with his front fins to keep them well supplied with oxygen. If the nest gets damaged, he repairs it.

The young sticklebacks hatch after about a week. They leave the nest but stay close by for another few weeks. If one of them wanders off on its own and is in danger of getting lost, its father quickly swims after it and brings it back.

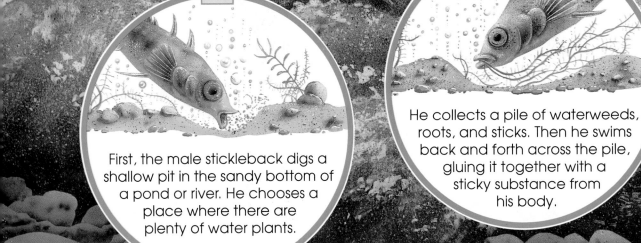

1 First, the male stickleback digs a shallow pit in the sandy bottom of a pond or river. He chooses a place where there are plenty of water plants.

2 He collects a pile of waterweeds, roots, and sticks. Then he swims back and forth across the pile, gluing it together with a sticky substance from his body.

3

Finally, the male stickleback bores
a hole in either side of the nest
to form a tunnel. He uses his
head to push through
the underwater nest.

Fun Quiz Time

Now that you have learned how various wild animals build their homes, why not test your knowledge with this fun quiz? The questions are all based on information given in the previous pages. And just in case you can't remember an answer, we've put all the answers on page 31. Good luck!

(a) Where does bees' wax come from?
(b) Why do the worker bees cluster together before they start building?
(c) What shape are the cells in a beehive?
(d) What do bees keep in the cells of their hive?

(a) Does the male or female weaverbird build the nest?
(b) What does a weaverbird build its nest around?
(c) Why do weaverbirds build long, narrow entrances to their nests?
(d) How many weaverbird nests might a single tree hold?

(a) What is a beaver's home called?
(b) How do beavers carry branches, twigs, and stones underwater?
(c) What material do beavers use to make their homes warm and watertight?
(d) How does a beaver reach its living chamber?

(a) What do spiders hope to catch in their webs?
(b) Is spider's silk weaker or stronger than steel wire?
(c) A spider spins two different types of silk. How are they different?
(d) How often does a spider have to build itself a new web?

(a) What do harvest mice use their tails for?
(b) Why do they build their nursery nest high above the ground?
(c) What tools does a harvest mouse use to shred and weave the grass for its nest?
(d) What does a harvest mouse store in its winter nest?

6

(a) What do tailor ants build their nests out of?
(b) What do tailor ants use instead of needles and thread?
(c) Which part of the larva's body produces silk?
(d) What happens if the gap between two leaves is too wide for one ant to cross?

7

(a) What tools does a mole use for digging?
(b) How is a mole's fur well suited to life underground?
(c) What would you find underneath the largest molehills?
(d) What sort of food does a mole keep in its larder?

8

(a) Why do male bowerbirds build their bowers?
(b) How does a bowerbird decorate its bower?
(c) Does the female lay her eggs in the bower itself or in another nest?
(d) What does a bowerbird use instead of a paintbrush?

9

(a) Where do chimpanzees build their sleeping nests?
(b) How often does a chimp build a new nest, and how long does it take?
(c) What does a chimp use its feet for as it builds its nest?
(d) How do young chimps practice making their own sleeping nests?

10

(a) What shape is a potter wasp's nest?
(b) How does a potter wasp soften the mud if it is too dry?
(c) What does a potter wasp put inside her nest for the wasp larva to feed on?
(d) With what does the potter wasp seal her nest?

11

(a) Does the male or the female stickleback build the underwater nest?
(b) What does the stickleback use to glue the nest together?
(c) How does the stickleback bore a tunnel through its nest?
(d) Why does a stickleback fan its eggs with fresh water?

GLOSSARY

Abdomen The tail end of an insect's or spider's body.

Architect A person who designs and plans buildings.

Beehive The name given to a bee's home. It is made up of rows of honeycombs.

Bower A structure of twigs built by the male bowerbird to attract a female.

Cell A six-sided wax chamber inside a beehive; also, the tiny single unit from which all living things are made.

Chamber A room or burrow.

Colony A group of animals or plants that live close together; for example, bees or ants.

Dam A barrier built to stop the flow of a river or stream.

Fertilize The combination of a substance (called sperm) from a male animal and a female's egg that results in the development of a new baby.

Gland Part of an animal's body that produces a special substance such as wax or silk.

Honeycomb A wax structure, made up of rows of six-sided cells, that hangs inside a beehive.

Larder A place used to store food.

Larva The stage in an insect's life between egg and pupa.

Litter Babies born at one time. Some animals have several groups of young, or litters, each year.

Lodge A beaver's home, built in a river or stream out of logs, branches, stones, and mud.

Molehill A pile of soil on the surface of the ground, pushed up by a mole as it digs its underground network of tunnels.

Oxygen A gas in the air and in water that animals and plants breathe in order to survive.

Parallel Two (or more) lines of things that lie next to each other, always at the same distance apart.

Pollen The powdery yellow dust made by the male part of a flower. It is eaten by bees and their larvae.

Queen bee The leader of a colony of bees. The queen is the only bee in the colony that is able to lay eggs.

Rodents The group of animals that includes mice, rats, beavers, and squirrels.

Scaffolding A structure put up to provide temporary support.

Spinneret A tiny nozzle on a spider's body out of which silk is squeezed.

Spokes The straight, supporting lines of silk in a spider's web.

Territory A large area in which an animal lives and searches for food.